GUT HEALTH AND PROBIOTICS

Gut Health Revolution: Comprehensive Guide To Probiotics, Nutrition, And Healthy Living For Enhanced Well-Being

Charlotte Harry

Table of Contents

CHAPTER ONE 4
- INTRODUCTION TO GUT HEALTH 4
- ANATOMY OF THE GUT 13
- GUT MICROBIOME 17
- GUT-BRAIN CONNECTION 23

CHAPTER TWO 28
- THE ROLE OF DIET IN GUT HEALTH 28
- NUTRIENTS AND DIGESTION 32
- FIBER AND PREBIOTICS 37
- FOODS TO SUPPORT GUT HEALTH 42
- DIETARY HABITS AND GUT HEALTH 47

CHAPTER THREE 52
- UNDERSTANDING PROBIOTICS 52
- WHAT ARE PROBIOTICS? 56
- SOURCES OF PROBIOTICS 60
- PROBIOTICS VS. PREBIOTICS 64
- MECHANISMS OF ACTION 69

CHAPTER FOUR 74
- BENEFITS OF PROBIOTICS FOR GUT HEALTH 74

CHAPTER FIVE 95
- CHOOSING AND USING PROBIOTICS 95

CHAPTER SIX 115
- RESEARCH AND ADVANCES IN PROBIOTICS 115
- CURRENT RESEARCH 118

FUTURE DIRECTIONS 123
PERSONALIZED NUTRITION AND PROBIOTICS 128
PROBIOTICS IN CLINICAL PRACTICE 133
CHAPTER SEVEN 139
LIFESTYLE FACTORS AND GUT HEALTH 139
THE END 162

CHAPTER ONE
INTRODUCTION TO GUT HEALTH

Gut health refers to the balance of microorganisms that live in the digestive tract. Maintaining the right balance of these microorganisms is crucial for physical and mental well-being. The gut, also known as the gastrointestinal (GI) tract, plays a key role in digesting food, absorbing nutrients, and eliminating waste. Additionally, the gut is involved in various bodily functions including metabolism, the immune system, and even brain health.

Good gut health means having a diverse community of bacteria and other microorganisms, often referred to as the gut microbiome. This diversity helps protect against harmful bacteria and supports the body's immune system. Poor gut health can lead to a range of problems, from digestive issues like bloating and constipation to more serious conditions like irritable bowel syndrome (IBS) and inflammatory bowel disease (IBD). Furthermore, an unhealthy gut can also contribute to weight gain, fatigue, and even mental health issues like anxiety and depression. Here's a straightforward guide to understanding gut health and why it matters.

Why Gut Health Matters

1. Digestion: A healthy gut ensures efficient digestion and absorption of nutrients. The good bacteria in your gut help break down food, making it easier for your body to absorb essential nutrients.

2. Immune System: About 70% of your immune system is housed in your gut. A healthy gut microbiome helps maintain a strong immune system by communicating with immune cells and fighting off infections.

3. Mental Health: There's a strong connection between your gut and your brain, known as the gut-brain axis. The gut produces neurotransmitters like serotonin, which affects mood and

mental health. A balanced gut microbiome can help reduce symptoms of anxiety and depression.

4. Weight Management: Gut health plays a role in maintaining a healthy weight. An imbalanced gut microbiome can lead to weight gain and obesity by influencing how your body stores fat and how it uses the energy from food.

Signs of an Unhealthy Gut

Several signs can indicate an unhealthy gut, including:

1. Digestive issues like bloating, gas, diarrhea, or constipation.
2. Frequent infections or illnesses.
3. Unintentional weight changes.
4. Skin problems like eczema.

5. Difficulty sleeping or constant fatigue.

6. Food intolerances.

How to Improve Gut Health

Improving gut health involves making lifestyle and dietary changes. Here are some tips:

1. Eat a Balanced Diet: Include plenty of fruits, vegetables, whole grains, and lean proteins in your diet. These foods are high in fiber and nutrients that feed the good bacteria in your gut.

2. Probiotics and Prebiotics: Probiotics are live bacteria found in fermented foods like yogurt, sauerkraut, and kimchi. Prebiotics are fibers that feed these good bacteria, found in foods like bananas, onions, and garlic.

3. Stay Hydrated: Drinking plenty of water helps with digestion and keeps the gut lining healthy.

4. Exercise Regularly: Physical activity can help maintain a healthy gut by promoting the growth of beneficial bacteria.

5. Reduce Stress: High stress levels can negatively affect gut health. Practices like meditation, deep breathing, and yoga can help manage stress.

6. Avoid Unnecessary Antibiotics: While antibiotics can be necessary for fighting bacterial infections, they can also kill beneficial bacteria in your gut. Use them only when prescribed by a healthcare professional.

Components of Gut Health

1. Gut Microbiota: This is the collection of microorganisms in the intestines. A healthy gut microbiota is diverse, meaning it contains a wide variety of different species. This diversity helps ensure that the gut can effectively process a wide range of foods and resist harmful microbes.

2. Gut Barrier: The lining of the gut serves as a barrier to prevent harmful substances from entering the bloodstream. This barrier is selectively permeable, allowing nutrients to pass through while keeping out toxins and pathogens. Maintaining the integrity of this barrier is crucial for preventing inflammation and other health issues.

3. Digestive Function: Efficient digestion and nutrient absorption are fundamental aspects of gut health. Enzymes and acids in the stomach break down food, which is then further processed in the intestines. Proper digestive function ensures that the body gets the nutrients it needs to function optimally.

Factors Influencing Gut Health

Several factors can influence gut health, including diet, lifestyle, and genetics.

1. Diet: What we eat has a significant impact on our gut microbiota. Diets high in fiber, fruits, vegetables, and fermented foods support a healthy gut by promoting the growth of beneficial bacteria. On the other hand, diets high

in processed foods, sugar, and unhealthy fats can harm gut health.

2. Lifestyle: Stress, lack of sleep, and physical inactivity can negatively affect gut health. Chronic stress and sleep deprivation can disrupt the balance of gut bacteria, while regular physical activity has been shown to promote a healthy gut microbiota.

3. Antibiotics and Medications: Antibiotics, while necessary for treating bacterial infections, can also kill beneficial bacteria in the gut. Other medications, such as nonsteroidal anti-inflammatory drugs (NSAIDs) and proton pump inhibitors (PPIs), can also affect gut health.

Importance of Gut Health

Maintaining a healthy gut is essential for overall well-being. A healthy gut supports efficient digestion, strong immune function, and balanced mental health. Emerging research suggests that an unhealthy gut may be linked to a variety of conditions, including obesity, diabetes, heart disease, and mental health disorders such as anxiety and depression.☐

ANATOMY OF THE GUT

The digestive system is a complex network of organs and glands essential for processing food and absorbing nutrients. This system includes several primary components: the mouth, esophagus, stomach, small intestine, large intestine (colon), rectum, and anus. Here, we will focus on the stomach, small intestine, and large intestine, as they are critical to gut health and overall digestion.

The stomach is a muscular organ where significant digestive processes begin. Upon food entry, the stomach secretes gastric acid and digestive enzymes that initiate the breakdown of food. This environment is highly acidic, facilitating the denaturation of proteins and the activation of digestive enzymes like pepsin. The stomach also performs a churning action, mechanically mixing the food with these digestive juices, transforming it into a semi-liquid substance known as chyme. This chyme is then gradually released into the small intestine for further digestion.

The small intestine is a lengthy, coiled tube where the majority of digestion and nutrient absorption occurs. It is divided into three parts: the duodenum, jejunum, and ileum. The duodenum receives the chyme from the stomach and continues the digestive process with enzymes from the pancreas and bile from the liver. The jejunum and ileum primarily handle nutrient absorption. The interior walls of the small intestine are lined with tiny, finger-like projections called villi, which greatly increase the surface area available for absorption. Each villus is covered with even smaller hair-like structures called microvilli, forming the brush border, which contains enzymes crucial for the final stages of nutrient digestion. Nutrients pass through the walls of the small intestine into the bloodstream, where they are transported to cells throughout the body.

The large intestine, or colon, plays a vital role in absorbing water and electrolytes from indigestible food residues, forming and excreting feces. The large intestine begins with the cecum, continues through the colon, and ends with the rectum and anus. The colon itself is subdivided into the ascending, transverse, descending, and sigmoid regions. A significant aspect of the large intestine is its dense population of bacteria, collectively known as the gut microbiota. These bacteria are essential for fermenting undigested carbohydrates, synthesizing certain vitamins, and forming short-chain fatty acids, which are beneficial for colon health. The final products of digestion are compacted into feces, stored in the rectum, and eventually expelled through the anus.

GUT MICROBIOME

The gut microbiome is a complex community of trillions of microorganisms that reside in our intestines. These microorganisms include bacteria, viruses, fungi, and other microbes, playing a crucial role in maintaining our overall health. Among these, bacteria are the most abundant and diverse, with the majority being beneficial and essential for healthy digestion. These helpful bacteria assist in breaking down food particles that the stomach and small intestine cannot fully digest. In doing so, they aid in nutrient absorption and the synthesis of essential vitamins, such as B vitamins and vitamin K.

A healthy gut microbiome is characterized by its diversity and balance. This means having a wide variety of microbial species coexisting and functioning harmoniously. Such diversity is critical because different microbes perform different functions, and their collective activity contributes to a robust digestive system and overall health. For instance, some bacteria are involved in fermenting dietary fibers, producing short-chain fatty acids that are important for colon health and have anti-inflammatory properties. Others play a role in modulating the immune system, protecting against pathogenic bacteria, and maintaining the integrity of the gut lining.

However, the delicate balance of the gut microbiome can be disrupted, leading to a state known as dysbiosis. Dysbiosis can result from various factors, including a poor diet high in processed foods and low in fiber, the use of antibiotics, which can kill beneficial bacteria along with harmful ones, illness, and chronic stress. When the microbiome is out of balance, it can have far-reaching effects on health.

Dysbiosis has been linked to a range of health issues. Digestive disorders such as irritable bowel syndrome (IBS) and inflammatory bowel diseases (IBD) like Crohn's disease and ulcerative colitis are often associated with an imbalanced gut microbiome. Beyond digestive health, dysbiosis has been implicated in metabolic conditions such as obesity and diabetes. For instance, certain microbial populations can influence how we metabolize food and store fat, thereby affecting body weight and insulin sensitivity.

Moreover, the gut-brain axis, a

bidirectional communication pathway between the gut and the brain, underscores the connection between gut health and mental health. Dysbiosis can contribute to mental health issues such as anxiety and depression. This is partly because the gut microbiome produces neurotransmitters like serotonin, which regulate mood, and partly because it modulates the body's stress response.

In addition to these conditions, dysbiosis has been linked to the development of allergies and autoimmune diseases. An imbalanced microbiome can lead to an overactive immune response, contributing to conditions such as asthma, eczema, and other allergic reactions. Similarly, a dysfunctional microbiome may play a role in autoimmune diseases by affecting the regulation of immune responses.

Maintaining a healthy gut microbiome involves adopting a balanced diet rich in diverse, fiber-containing foods, minimizing unnecessary antibiotic use, managing stress, and leading a healthy lifestyle. Probiotics and prebiotics can also support gut health by promoting the growth of beneficial bacteria. Probiotics are live beneficial bacteria found in fermented foods like yogurt and sauerkraut, while prebiotics are non-digestible fibers that feed these beneficial bacteria.

GUT-BRAIN CONNECTION

The gut-brain connection, commonly referred to as the gut-brain axis, is a complex communication network linking the gut and the brain. This intricate system involves both physical and biochemical pathways, primarily through the nervous system and the endocrine system. The vagus nerve, a critical component of the parasympathetic nervous system, plays a pivotal role in this communication. It acts as a conduit, transmitting signals between the gut and the brain, thereby influencing various physiological processes.

One of the remarkable aspects of the gut-brain axis is the production of neurotransmitters within the digestive system. Neurotransmitters are chemical messengers that transmit signals across nerve cells, influencing numerous bodily functions and emotional states. Serotonin, a neurotransmitter widely recognized for its role in mood regulation, is predominantly produced in the gut. In fact, about 90% of the body's serotonin is synthesized in the digestive tract. This significant production of serotonin in the gut underscores the profound impact gut health can have on emotional well-being and cognitive function.

Furthermore, the gut microbiome, which consists of trillions of microorganisms residing in the digestive tract, plays an influential role in the gut-brain connection. These microorganisms interact with the central nervous system, thereby affecting brain chemistry and behavior. A balanced gut microbiome contributes to overall mental health, potentially reducing the risk of conditions such as depression and anxiety. This is because the gut microbiota can modulate the production of neurotransmitters and inflammatory markers that are critical for brain function and mood regulation.

Conversely, an unhealthy gut can have detrimental effects on mental health. Dysbiosis, an imbalance in the gut microbiome, has been associated with increased risks of mental health disorders. This imbalance can lead to the production of harmful substances that affect brain function, contributing to conditions such as anxiety and depression. Additionally, chronic inflammation stemming from gut issues can further exacerbate these mental health problems.

Research continues to unveil the intricate ways in which gut health impacts the brain. For instance, studies have shown that probiotics, which are beneficial bacteria that help maintain a healthy gut microbiome, can have positive effects on mood and cognitive functions. These findings highlight the potential for dietary interventions and probiotic supplements to support mental health through the modulation of gut bacteria.

CHAPTER TWO
THE ROLE OF DIET IN GUT HEALTH

The human gut, home to trillions of microorganisms, plays a crucial role in overall health. These microorganisms, collectively known as the gut microbiota, aid in digestion, bolster the immune system, and even influence mood and mental health. Diet significantly impacts the composition and function of the gut microbiota, making it a pivotal factor in maintaining gut health.

1. Fiber-Rich Foods: Fiber is a type of carbohydrate that the body cannot digest. Instead, it passes through the digestive system relatively intact, providing various benefits along the way. Foods rich in fiber, such as fruits,

vegetables, legumes, and whole grains, serve as food for the gut bacteria. When gut bacteria ferment fiber, they produce short-chain fatty acids (SCFAs), which have anti-inflammatory properties and support the gut lining. Moreover, a high-fiber diet promotes the growth of beneficial bacteria, enhancing gut diversity and stability.

2. Fermented Foods: Fermented foods, such as yogurt, kefir, sauerkraut, kimchi, and kombucha, are rich in probiotics—live beneficial bacteria. Consuming these foods can introduce more good bacteria into the gut, helping to maintain a balanced microbial environment. This balance is essential for preventing the overgrowth of harmful bacteria that can

lead to digestive issues and other health problems.

3. Prebiotic Foods: Prebiotics are compounds in food that induce the growth or activity of beneficial microorganisms. They are primarily found in fiber-rich foods but can also be present in garlic, onions, leeks, asparagus, bananas, and artichokes. Prebiotics enhance gut health by feeding the beneficial bacteria and promoting their growth and activity.

4. Polyphenol-Rich Foods: Polyphenols are plant compounds found in foods like berries, nuts, seeds, green tea, coffee, and dark chocolate. They have antioxidant properties and are metabolized by gut bacteria into

beneficial compounds. Polyphenols support the growth of good bacteria while inhibiting the growth of harmful ones.

5. Avoiding Harmful Foods: Certain foods can negatively impact gut health. Diets high in refined sugars and unhealthy fats can disrupt the balance of gut bacteria, leading to inflammation and a weakened gut lining. Additionally, excessive consumption of artificial sweeteners, preservatives, and processed foods can harm the gut microbiota.

6. Hydration: Adequate hydration is essential for digestion and maintaining a healthy gut lining. Water helps dissolve nutrients and waste products, making it

easier for them to pass through the gut. It also ensures that the digestive system functions smoothly.

7. Moderation and Variety: A varied diet ensures a diverse gut microbiota, which is associated with better health outcomes. Consuming a wide range of foods provides different nutrients and types of fiber, supporting the growth of various beneficial bacteria.

NUTRIENTS AND DIGESTION

The process of digestion, a complex series of events that convert food into nutrients the body can use, begins as soon as food enters the mouth and continues as it traverses through the stomach, small intestine, and large intestine. Each class of nutrients undergoes distinct processing stages, reflecting their unique chemical structures and roles in the body.

Carbohydrates, the body's primary energy source, begin their breakdown in the mouth. Saliva contains enzymes, notably amylase, which initiate the conversion of complex carbohydrates into simpler sugars. As the partially digested carbohydrates move into the stomach, the acidic environment temporarily halts carbohydrate digestion. However, the process resumes with greater intensity in the small intestine. Here, pancreatic amylase continues breaking down complex carbohydrates into simple sugars like glucose, which are then absorbed into the bloodstream through the intestinal walls. This absorption primarily occurs in the jejunum and ileum, parts of the small intestine designed to maximize nutrient uptake.

Proteins, essential for growth and repair, begin their digestive journey in the stomach. Gastric juices, rich in hydrochloric acid and the enzyme pepsin, denature proteins and break them into smaller polypeptides and amino acids. This acidic environment is crucial for pepsin's activity and for breaking down protein structures. The partially digested proteins then enter the small intestine, where pancreatic enzymes, including trypsin and chymotrypsin, further degrade them into amino acids. These amino acids are absorbed through the intestinal lining and transported to the liver and other parts of the body for synthesis and energy.

Fats, vital for energy storage and cellular structure, undergo a unique digestion process primarily in the small intestine. The liver produces bile, stored in the gallbladder and released into the small intestine when fat is present. Bile acts as an emulsifier, breaking down large fat globules into smaller droplets, increasing their surface area for enzyme action. Pancreatic lipase then efficiently breaks these droplets into fatty acids and glycerol. These smaller molecules are absorbed by the enterocytes lining the small intestine and are transported via the lymphatic system into the bloodstream.

Vitamins and minerals, essential for various bodily functions, do not require breakdown and are absorbed directly. Fat-soluble vitamins (A, D, E, and K) are absorbed alongside dietary fats, while water-soluble vitamins (B and C) are absorbed directly into the bloodstream. The small intestine, particularly the duodenum and jejunum, is the primary site for mineral absorption, although some minerals are absorbed in the large intestine.

Thus, digestion is a highly coordinated process, with each nutrient type undergoing specific breakdown and absorption pathways, ensuring the body receives the necessary components for energy, growth, and overall function.

FIBER AND PREBIOTICS

Dietary fiber and prebiotics are essential components of a healthy diet, playing vital roles in maintaining gut health and overall well-being.

Dietary Fiber

Dietary fiber is a type of carbohydrate found in plant-based foods that the body cannot digest. It comes in two forms: soluble and insoluble, each with distinct benefits for digestive health.

1. Soluble Fiber: This type of fiber dissolves in water to form a gel-like substance. Soluble fiber is beneficial for lowering blood cholesterol and glucose

levels, making it a key component in managing heart health and diabetes. Foods rich in soluble fiber include oats, peas, beans, apples, citrus fruits, carrots, barley, and psyllium. By slowing digestion, soluble fiber helps to regulate blood sugar levels and provides a feeling of fullness, which can aid in weight management.

2. Insoluble Fiber: Unlike soluble fiber, insoluble fiber does not dissolve in water. Instead, it adds bulk to the stool, which helps food pass more quickly through the stomach and intestines. This type of fiber is crucial for preventing constipation and promoting regular bowel movements. It is found in foods such as whole wheat flour, wheat

bran, nuts, beans, and vegetables like cauliflower, green beans, and potatoes. Insoluble fiber helps maintain bowel health and regularity, reducing the risk of developing hemorrhoids and diverticular disease.

Both types of fiber are important for a healthy digestive system. They work together to ensure that food moves smoothly through the digestive tract, preventing issues such as constipation, bloating, and gas. A diet high in fiber is also associated with a lower risk of developing chronic diseases, including heart disease, stroke, type 2 diabetes, and certain types of cancer.

Prebiotics

Prebiotics are a specific type of non-digestible fiber that serve as food for the beneficial bacteria in the gut. Unlike probiotics, which are live bacteria found in fermented foods and supplements, prebiotics help nourish and increase the population of these healthy bacteria. This process is crucial for maintaining a balanced gut microbiome, which is essential for digestion, immunity, and overall health.

Foods rich in prebiotics include garlic, onions, leeks, asparagus, bananas, chicory root, and whole grains. When consumed, prebiotics pass through the upper part of the gastrointestinal tract undigested and reach the colon, where they are fermented by the gut microbiota. This fermentation process produces short-chain fatty acids, which provide numerous health benefits, including enhanced mineral absorption, reduced inflammation, and improved gut barrier function.

By increasing the population of beneficial bacteria, prebiotics help to outcompete harmful bacteria, reducing the risk of infections and promoting a healthy gut environment. They also play a role in enhancing immune function, as a significant portion of the immune system is located in the gut. Additionally, a healthy gut microbiome is linked to improved mental health, as it communicates with the brain through the gut-brain axis.

Incorporating a variety of high-fiber foods and prebiotics into your diet is essential for supporting digestive health, boosting immunity, and enhancing overall well-being. Together, dietary fiber and prebiotics create a foundation for a healthy gut, promoting optimal digestion and protecting against various diseases.

FOODS TO SUPPORT GUT HEALTH

Maintaining good gut health is crucial for overall well-being, as the gut plays a significant role in digestion, nutrient absorption, and immune function. A diet rich in specific foods can support and enhance gut health due to their nutrient content, fiber, prebiotics, and probiotics. Here are some key food categories that are particularly beneficial for gut health:

Fruits: Incorporating a variety of fruits into your diet can significantly improve gut health. Apples, for instance, are rich in fiber, particularly pectin, which acts as a prebiotic to feed beneficial gut bacteria. Bananas are another excellent choice as they provide a blend of fiber, vitamins, and minerals like potassium, which support digestive health. Berries, such as blueberries, strawberries, and raspberries, are packed with antioxidants and fiber, aiding in reducing inflammation and promoting a healthy gut lining. Citrus fruits like oranges and grapefruits are also beneficial due to their high vitamin C content and soluble fiber.

Vegetables: Consuming a wide range of vegetables is essential for gut health. Leafy greens such as spinach, kale, and Swiss chard are high in fiber and contain essential vitamins and minerals that promote a healthy gut microbiome. Cruciferous vegetables like broccoli, Brussels sprouts, and cauliflower are rich in fiber and sulfur-containing compounds that support the growth of beneficial bacteria. Root vegetables such as carrots, sweet potatoes, and beets are also excellent sources of fiber and nutrients, supporting overall digestive health.

Whole Grains: Whole grains are a significant source of dietary fiber, which is crucial for maintaining regular bowel movements and supporting a healthy gut microbiota. Oats, for example, contain beta-glucan, a type of soluble fiber that promotes the growth of beneficial bacteria. Brown rice, quinoa, and barley are other excellent whole grain options that provide fiber and essential nutrients, contributing to a well-functioning digestive system.

Fermented Foods: Fermented foods are a powerhouse for gut health due to their probiotic content. Probiotics are live beneficial bacteria that can help balance the gut microbiome. Yogurt, especially those with live active cultures, is a well-known source of probiotics. Kefir, a fermented milk drink, contains a diverse range of probiotic strains. Fermented vegetables like sauerkraut and kimchi are rich in probiotics and also provide fiber and vitamins. Kombucha, a fermented tea, offers probiotics and antioxidants, contributing to a healthy gut environment.

Legumes: Beans, lentils, and chickpeas are legumes that are high in fiber and protein, both of which aid in digestion and provide essential nutrients. The fiber in legumes acts as a prebiotic, feeding beneficial gut bacteria and promoting a balanced gut microbiome. Additionally, legumes contain resistant starch, which is not digested in the small intestine but ferments in the colon, producing beneficial short-chain fatty acids that support gut health.

Incorporating these foods into your daily diet can help maintain and improve gut health, supporting overall digestive function, enhancing nutrient absorption, and bolstering the immune system. By focusing on a diverse and balanced intake of fruits, vegetables, whole grains, fermented foods, and legumes, you can create a gut-friendly diet that promotes long-term health.

DIETARY HABITS AND GUT HEALTH

The way and timing of your eating habits can profoundly impact your gut health. Implementing certain dietary practices can enhance digestion, prevent common gastrointestinal issues, and promote overall well-being. Here are some key habits to consider for optimal gut function:

1. Regular Meal Timing: Maintaining consistent meal times helps regulate your digestive system, minimizing problems like acid reflux and indigestion. Eating at irregular intervals can disrupt your body's natural rhythms, leading to digestive discomfort. Aim for balanced meals and snacks spread evenly throughout the day, avoiding erratic eating patterns. This consistency helps your body

anticipate and prepare for food intake, ensuring smoother digestion.

2. Chewing Thoroughly: Proper chewing is essential for breaking down food into smaller, more manageable pieces. This not only eases the workload on your stomach but also enhances nutrient absorption. Chewing thoroughly initiates the digestive process in the mouth, where enzymes begin breaking down food components, making it easier for your stomach and intestines to handle. By taking the time to chew well, you can improve your digestion and nutrient uptake significantly.

3. Hydration: Adequate hydration is crucial for effective digestion. Water aids

in breaking down food, allowing your body to absorb nutrients more efficiently. It also helps soften stool, preventing constipation. Aim to drink at least eight glasses of water a day, increasing this amount if you are physically active. Staying well-hydrated supports the smooth movement of food through your digestive tract and maintains the health of your gut lining.

4. Mindful Eating: Practicing mindful eating involves paying close attention to what and how much you eat. This habit can prevent overeating and promote better digestion. Eating slowly and without distractions allows your body to signal when you are full, which can help you maintain a healthy weight and avoid

digestive discomfort. Mindful eating encourages you to savor your food, appreciate its flavors, and be more attuned to your body's hunger and satiety cues.

5. Avoiding Overly Processed Foods: Highly processed foods often contain unhealthy fats, sugars, and additives that can negatively affect gut bacteria and disrupt digestion. These foods can lead to inflammation and an imbalance in your gut microbiome. Instead, focus on consuming whole, minimally processed foods like fruits, vegetables, whole grains, and lean proteins. These nutrient-dense options support a healthy gut by providing essential vitamins, minerals, and fiber.

CHAPTER THREE

UNDERSTANDING PROBIOTICS

Probiotics are live microorganisms that provide health benefits when consumed in adequate amounts. Often referred to as "good" or "friendly" bacteria, they play a crucial role in maintaining a balanced gut microbiome, which is essential for overall health.

How Do Probiotics Work?

Probiotics contribute to health by balancing the bacteria in your gut. The human gut contains trillions of bacteria, some beneficial and others potentially harmful. An imbalance, where harmful bacteria outnumber the good ones, can lead to digestive issues, infections, and other health problems. By adding more good bacteria, probiotics help maintain a healthy balance.

These beneficial bacteria help digest food, absorb nutrients, and fend off harmful bacteria that cause infections. They also produce certain vitamins and aid in the fermentation of dietary fibers, turning them into short-chain fatty acids beneficial for gut health.

Benefits of Probiotics

1. Digestive Health: Probiotics can help prevent and treat diarrhea, especially when it's caused by antibiotics, which can kill both good and bad bacteria. They are also beneficial for managing conditions like irritable bowel syndrome (IBS) and inflammatory

bowel diseases such as Crohn's disease and ulcerative colitis.

2. Immune System Support: A healthy gut microbiome is vital for a robust immune system. Probiotics can enhance the production of natural antibodies and increase the activity of immune cells like T lymphocytes and natural killer cells.

3. Mental Health: There is a strong connection between gut health and mental health, often referred to as the gut-brain axis. Probiotics can help manage symptoms of anxiety, depression, and other mental health disorders.

4. Skin Health: Certain probiotics can improve skin conditions such as eczema

and acne. They work by reducing inflammation and promoting a balanced gut flora, which in turn reflects on the skin.

5. Heart Health: Some probiotics can help lower blood pressure and cholesterol levels, reducing the risk of heart disease.

Choosing the Right Probiotic

When selecting a probiotic, consider the specific strains included and their intended benefits. Different strains serve different purposes, so it's important to choose one that aligns with your health needs. Also, pay attention to the colony-forming units (CFUs) on the label, which indicate the number of live bacteria in each dose.

WHAT ARE PROBIOTICS?

Probiotics are live microorganisms that confer health benefits when consumed in adequate amounts. Often called "good" or "friendly" bacteria, probiotics help maintain the balance of beneficial bacteria in the gut. These beneficial bacteria are similar to the naturally occurring microorganisms in the digestive system and play a crucial role in restoring the natural balance of the gut microbiome, especially when it has been disrupted by factors such as antibiotics, illness, or an unhealthy diet.

The gut microbiome is a complex community of trillions of microorganisms, including bacteria, viruses, fungi, and other microbes, that reside in the digestive tract. A balanced gut microbiome is essential for various aspects of health, including digestion, nutrient absorption, immune function, and even mental health. When the balance of the gut microbiome is disrupted, it can lead to a range of health issues, from digestive disorders to weakened immunity and beyond. Probiotics can help restore this balance by replenishing the population of beneficial bacteria.

There are many types of probiotics, but the most common groups are Lactobacillus, Bifidobacterium, and Saccharomyces boulardii. Each group contains different strains that provide specific health benefits.

Lactobacillus: This is one of the most common types of probiotics and is found in yogurt and other fermented foods. Many strains of Lactobacillus can help with diarrhea and are particularly beneficial for people who have difficulty digesting lactose, the sugar found in milk. By producing lactase, the enzyme that breaks down lactose, Lactobacillus can help reduce symptoms of lactose intolerance. Additionally, some Lactobacillus strains can help prevent and treat infections, reduce inflammation, and boost the immune system.

Bifidobacterium: Commonly found in dairy products, Bifidobacterium strains are known for their ability to ease symptoms of irritable bowel syndrome (IBS) and other digestive conditions. These bacteria help maintain the integrity of the gut lining, prevent the growth of harmful bacteria, and enhance the overall function of the immune system. Bifidobacterium also aids in the digestion of dietary fiber, producing short-chain fatty acids that are beneficial for gut health.

Saccharomyces boulardii: Unlike Lactobacillus and Bifidobacterium, Saccharomyces boulardii is a type of yeast found in probiotics. It is particularly effective in combating diarrhea and other digestive problems. Saccharomyces boulardii works by enhancing the gut's immune response, inhibiting the growth of harmful bacteria and pathogens, and reducing inflammation. It is often used to prevent and treat diarrhea caused by antibiotics, infections, and other factors.

SOURCES OF PROBIOTICS

Probiotics, beneficial bacteria that support digestive health and overall well-being, can be obtained from both natural sources and dietary supplements. These sources offer diverse options to incorporate probiotics into one's diet conveniently.

Natural Sources of Probiotics

Many traditional fermented foods are teeming with probiotics, making them popular choices for natural intake:

1. Yogurt: This dairy product is created by fermenting milk with specific bacteria strains like Lactobacillus bulgaricus and Streptococcus thermophilus. It remains one of the most widely consumed probiotic foods globally due to its creamy texture and versatile flavor profiles.

2. Kefir: Originating from the Caucasus Mountains, kefir is a fermented milk drink typically cultured with kefir grains. It contains a diverse mix of bacteria and yeasts, offering a broader spectrum of probiotic strains compared to yogurt.

3. Sauerkraut: This German staple is made from fermented cabbage, rich in probiotics like Lactobacillus bacteria. It

provides a tangy flavor and crunchy texture, often used in sandwiches or as a side dish.

4. Kimchi: A staple of Korean cuisine, kimchi is fermented vegetables (commonly cabbage and radishes) seasoned with spices like garlic and chili peppers. It offers a spicy, pungent flavor profile and is a potent source of probiotics.

5. Miso: A traditional Japanese seasoning made by fermenting soybeans with salt and a type of fungus called koji. Miso paste is used to make soups, marinades, and dressings, contributing a savory umami taste along with probiotics.

6. Tempeh: Originating from Indonesia, tempeh is a fermented soybean product bound together into a dense cake. It provides a nutty flavor and is a good source of probiotics along with protein and dietary fiber.

7. Kombucha: This effervescent tea beverage is made by fermenting sweetened black or green tea with a symbiotic culture of bacteria and yeast (SCOBY). Kombucha offers a slightly tangy taste and is enjoyed chilled as a refreshing probiotic drink.

Supplements:

In addition to natural sources, probiotics are available as dietary supplements in various forms such as capsules, tablets, powders, and liquids.

These supplements provide concentrated doses of specific probiotic strains, offering a convenient option for individuals who may not consume fermented foods regularly or who seek targeted probiotic benefits. Supplements are particularly useful for maintaining gut health, supporting immune function, and restoring microbial balance after antibiotic use.

PROBIOTICS VS. PREBIOTICS

Probiotics and prebiotics play crucial roles in maintaining a healthy gut microbiome, though they differ significantly in their functions and forms within the body.

Probiotics are living microorganisms, primarily bacteria, that confer health benefits when consumed in adequate amounts. They contribute to the population of beneficial bacteria already present in the digestive system. By doing so, probiotics help restore and maintain a balanced gut microbiome, which is essential for efficient digestion and overall health. These microorganisms can be found in various fermented foods such as yogurt, kefir, sauerkraut, and kimchi, as well as in dietary supplements. Probiotics are known for their ability to alleviate digestive issues such as diarrhea, bloating, and irritable bowel syndrome (IBS). They may also boost the immune system and potentially aid in managing conditions like eczema and urinary tract infections.

On the other hand, prebiotics are non-digestible fibers that serve as fuel for probiotics. Unlike probiotics, which are live organisms, prebiotics themselves are not alive. Instead, they pass through the digestive tract undigested until they reach the colon, where they are fermented by the beneficial bacteria residing there. This fermentation process produces short-chain fatty acids, which provide energy for the cells lining the colon and contribute to overall gut health. Foods rich in prebiotics include garlic, onions, leeks, asparagus, bananas, and whole grains like oats and barley. Including these foods in your diet can help stimulate the growth and activity of probiotics, thereby enhancing their beneficial effects on gut health.

The synergy between probiotics and prebiotics is crucial for optimizing gut health. While probiotics introduce beneficial bacteria into the gut microbiome, prebiotics ensure that these bacteria have the nutrients they need to thrive and perform their functions effectively. This cooperative relationship results in a healthier balance of gut flora, which in turn supports better digestion, nutrient absorption, and immune function. Moreover, a diverse and balanced gut microbiome has been linked to numerous health benefits beyond digestion, including mental health and even weight management.

Incorporating both probiotics and prebiotics into your diet through food sources or supplements can be a proactive way to support your gut health. However, it's essential to note that individual responses to probiotics and prebiotics may vary, and consulting with a healthcare professional can help determine the best approach for your specific needs. By nurturing your gut microbiome with these beneficial components, you can promote overall well-being and enhance your body's natural defenses against various health challenges.

MECHANISMS OF ACTION

Probiotics exert numerous beneficial effects on the gut microbiome through various mechanisms, each contributing to overall digestive health and immune function.

Firstly, probiotics play a pivotal role in restoring microbial balance within the gut. This balance between beneficial and harmful bacteria is crucial for optimal digestive function and overall well-being. When this balance is disrupted, it can lead to digestive disorders and susceptibility to infections. Probiotics help in restoring this equilibrium by replenishing populations of beneficial bacteria, thereby promoting a healthier gut environment.

Secondly, probiotics enhance the integrity of the gut barrier, a vital defense mechanism against pathogens and toxins. The gut barrier consists of a mucosal layer and tight junction proteins that regulate the passage of molecules into the bloodstream. Probiotics stimulate the production of mucus and strengthen tight junctions, thereby fortifying the gut lining and reducing the permeability that can lead to inflammation and disease.

Additionally, probiotics compete effectively with pathogenic bacteria for nutrients and attachment sites in the gut. By outcompeting these harmful microbes, probiotics help prevent their colonization and reduce the risk of infections. This competitive exclusion mechanism is crucial for maintaining a balanced and healthy gut microbiome.

Moreover, certain strains of probiotics produce antimicrobial substances such as lactic acid, hydrogen peroxide, and bacteriocins. These substances inhibit the growth of pathogenic bacteria, further contributing to a hostile environment for potential invaders while supporting the dominance of beneficial microbes.

Beyond their direct effects on gut flora, probiotics interact with the gut-associated lymphoid tissue (GALT), a significant component of the immune system. Probiotics influence immune responses by modulating the activity of immune cells and enhancing the production of antibodies. This modulation helps in strengthening immune defenses against infections and other immune-related disorders.

Thua, probiotics provide multifaceted benefits to the gut microbiome by restoring microbial balance, enhancing gut barrier function, competing with pathogens, producing antimicrobial substances, and modulating immune responses. These mechanisms collectively contribute to improved digestive health, reduced risk of infections, and enhanced overall well-being. Incorporating probiotics into daily dietary practices or as supplements can therefore play a valuable role in maintaining gut health and supporting immune function.

CHAPTER FOUR
BENEFITS OF PROBIOTICS FOR GUT HEALTH

Probiotics have gained significant attention for their potential health benefits, particularly in supporting gut health. Here's a breakdown of how probiotics contribute to various aspects of well-being:

Digestive Health

Probiotics play a crucial role in maintaining digestive health, offering effective solutions to a variety of gastrointestinal issues. Among their primary benefits is their ability to alleviate symptoms associated with conditions like Irritable Bowel Syndrome (IBS), constipation, and diarrhea.

In the case of Irritable Bowel Syndrome, probiotics, particularly strains such as Bifidobacterium and Lactobacillus, have demonstrated significant promise. Symptoms like abdominal pain, bloating, and irregular bowel movements can be mitigated by these beneficial bacteria. They achieve this by regulating gut motility and reducing inflammation within the intestines, thereby providing relief and improving overall digestive function.

For individuals struggling with constipation, specific probiotic strains can prove beneficial. These strains work by softening stool and enhancing bowel movements, making it easier to pass stools regularly. By increasing the bulk and moisture content of stool and improving the contraction of intestinal muscles, probiotics facilitate smoother digestion and relieve discomfort associated with constipation.

Perhaps one of the most well-known benefits of probiotics is their effectiveness in preventing and treating various forms of diarrhea. They are particularly valuable in cases of antibiotic-associated diarrhea and infectious diarrhea caused by pathogens like Clostridium difficile or viruses. Probiotics help restore the balance of gut bacteria that is often disrupted by antibiotics or infections. This restoration not only reduces the severity and duration of diarrhea episodes but also supports overall gastrointestinal health.

The mechanisms through which probiotics exert their beneficial effects are multifaceted. They compete with harmful bacteria for nutrients and attachment sites within the gut, thereby inhibiting the growth of pathogens. Additionally, probiotics produce antimicrobial substances that directly target harmful microbes, further promoting a healthier microbial balance. Moreover, these beneficial bacteria modulate the immune response in the intestinal lining, enhancing its ability to defend against infections and inflammatory responses.

By fostering a healthier gut environment, probiotics contribute significantly to alleviating digestive discomfort and improving overall well-being. Their ability to regulate gut motility, reduce inflammation, and restore microbial balance makes them a valuable therapeutic option for individuals suffering from various digestive ailments.

Thus, probiotics represent a natural and effective approach to maintaining digestive health. Whether addressing symptoms of IBS, constipation, or diarrhea, these beneficial bacteria offer solutions that are not only symptom-relieving but also supportive of long-term gastrointestinal wellness. As our understanding of probiotics continues to evolve, so too does their potential to provide personalized and targeted relief for individuals seeking to optimize their digestive function and overall quality of life.

Immune System Support

The gut serves as a vital cornerstone of the body's immune system, hosting a substantial portion of its immune cells and playing a critical role in immune function. Among the factors influencing this intricate relationship are probiotics, which exert multifaceted effects on immune responses.

Enhancing Gut Barrier Function

Probiotics contribute significantly to reinforcing the intestinal barrier, a crucial defense mechanism against pathogens and toxins. They facilitate the production of mucin, a key component of mucus that lines the intestinal walls. This mucus acts as a physical barrier, trapping potential invaders and preventing their entry into the bloodstream. Moreover, probiotics stimulate the synthesis of tight junction proteins, which are essential for maintaining the integrity of the gut epithelial barrier. By sealing the gaps between cells, these proteins further fortify the barrier, reducing the risk of infections and enhancing overall gut health.

Regulating Immune Responses

Within the gut-associated lymphoid tissue (GALT), probiotics engage directly with immune cells, influencing their behavior and function. This interaction extends to various aspects of immune regulation, including the modulation of cytokine production and the promotion of antibody responses. Cytokines are signaling molecules that coordinate immune responses, while antibodies are crucial for recognizing and neutralizing pathogens. By influencing these immune activities, probiotics help maintain a delicate balance within the immune system, ensuring it responds appropriately to challenges without overreacting.

Reducing Inflammation

Chronic inflammation within the gut mucosa is associated with numerous health issues, including autoimmune disorders. Probiotics exhibit notable anti-inflammatory properties that help mitigate this inflammation. By dampening the excessive immune response in the gut, probiotics contribute to a healthier environment within the intestines. This, in turn, supports overall immune health by preventing the development of inflammatory conditions that can compromise immune function.

Clinical Insights and Applications

Research into probiotics' immunomodulatory effects has led to various clinical applications. For instance, certain probiotic strains have shown efficacy in managing conditions such as inflammatory bowel disease (IBD), where immune dysregulation and inflammation play central roles in disease progression. By promoting gut barrier integrity and regulating immune responses, probiotics offer a promising avenue for adjunctive therapy in such chronic inflammatory conditions.

Considerations and Future Directions

While the potential benefits of probiotics on immune health are promising, it's essential to consider individual variability in gut microbiota composition and response to probiotic interventions. Factors such as diet, genetics, and environmental influences can all impact how probiotics interact with the immune system and overall health outcomes.

Looking forward, ongoing research aims to further elucidate the specific mechanisms by which probiotics influence immune function. This deeper understanding will not only enhance our knowledge of gut-immune interactions but also inform the development of more targeted probiotic therapies tailored to individual health needs.

However, probiotics play a pivotal role in supporting immune function through their actions on gut barrier function, immune response regulation, and inflammation modulation. By bolstering the gut's defenses and maintaining immune balance, probiotics contribute significantly to overall immune health and may offer therapeutic potential in managing immune-related disorders.

Mental Health

The gut-brain axis serves as a sophisticated bidirectional communication network linking the gut and the brain, facilitating a profound interaction between emotional and cognitive functions and peripheral intestinal activities. Probiotics, beneficial bacteria known for their health-promoting properties, have emerged as significant players in influencing mental health and mood through various mechanisms.

One pivotal way probiotics impact mental health is through the production of neurotransmitters. Strains such as Lactobacillus and Bifidobacterium are capable of synthesizing neurotransmitters like serotonin and gamma-aminobutyric acid (GABA) within the gut. Serotonin, often referred to as the "happy hormone," regulates mood and has a profound impact on emotional well-being. GABA, on the other hand, helps in reducing neuronal excitability and plays a crucial role in anxiety regulation. By producing these neurotransmitters, probiotics can directly influence mood, anxiety levels, and stress responses.

Another crucial mechanism involves the reduction of inflammation. Chronic inflammation in the gut has been linked to several mood disorders, including depression and anxiety. Probiotics contribute to gut health by enhancing intestinal barrier function and decreasing inflammation levels. By bolstering the gut barrier integrity, probiotics help prevent the leakage of harmful substances into the bloodstream, which can trigger systemic inflammation and affect brain function. This dual action of maintaining gut barrier integrity and reducing inflammation underscores probiotics' potential in alleviating symptoms associated with mood disorders.

Moreover, probiotics exert an impact on the stress response through modulation of the hypothalamic-pituitary-adrenal (HPA) axis. This axis regulates the body's reaction to stress, involving the release of cortisol and other stress hormones. Research suggests that probiotics can influence cortisol levels and the production of stress-related neurotransmitters, thereby mitigating the physiological effects of stress on mental well-being. By promoting a balanced stress response, probiotics contribute to resilience against stressors and support overall mental health.

Probiotics play a multifaceted role in mental health by producing neurotransmitters crucial for mood regulation, reducing gut inflammation linked to mood disorders, and modulating the body's stress response mechanisms. These mechanisms highlight the potential of probiotics as therapeutic agents in managing conditions like depression, anxiety, and stress-related disorders. As ongoing research continues to elucidate the intricate workings of the gut-brain axis, probiotics offer promising avenues for integrative approaches to mental health care, emphasizing the interconnectedness of gut health and emotional well-being.

Metabolic Health

Probiotics have garnered attention not only for their role in gut health but also for their potential impact on metabolic health, particularly in weight management and combating metabolic disorders.

In terms of weight management, specific strains of probiotics have shown promise in contributing to modest reductions in body weight and fat mass. These effects are believed to be mediated through several mechanisms. Probiotics can influence appetite-regulating hormones, such as leptin and ghrelin, which play crucial roles in controlling hunger and satiety. Additionally, certain probiotics increase the production of short-chain fatty acids (SCFAs) in the gut. SCFAs like acetate, propionate, and butyrate are known to promote fat metabolism and regulate energy balance, potentially aiding in weight loss efforts. Furthermore, probiotics have been observed to modulate the expression of genes involved in lipid metabolism, suggesting another avenue through which they may impact body composition.

In terms of blood sugar control, probiotics have shown promise in improving insulin sensitivity and glucose metabolism. By reducing systemic inflammation and enhancing the integrity of the gut barrier, probiotics may help mitigate the onset and progression of insulin resistance, a key precursor to type 2 diabetes. SCFAs produced by probiotics also play a role here, as they have been implicated in regulating glucose and lipid metabolism in the body.

Another area where probiotics demonstrate beneficial effects is in lipid profile improvement. Studies have indicated that certain probiotic strains can contribute to reducing levels of total cholesterol and LDL cholesterol (often referred to as "bad" cholesterol). These changes are significant as they contribute to better cardiovascular health outcomes, potentially lowering the risk of heart disease and related complications.

The mechanisms underlying these metabolic benefits are multifaceted. Probiotics interact with the gut microbiota, influencing its composition and metabolic activities. This interaction affects various physiological processes, including nutrient absorption, inflammation modulation, and metabolic signaling pathways. Moreover, probiotics' ability to enhance gut barrier function is crucial, as a compromised gut barrier (leaky gut) has been implicated in systemic inflammation and metabolic disorders.

While further research is needed to fully elucidate the specific strains and mechanisms involved, the emerging evidence suggests that probiotics play a significant role in promoting metabolic health. Their potential to aid in weight management, improve blood sugar control, and optimize lipid profiles underscores their importance as a complementary approach to managing metabolic disorders like obesity, insulin resistance, and cardiovascular disease. As our understanding deepens, incorporating probiotics into dietary strategies may become increasingly integral to comprehensive metabolic health management protocols.

CHAPTER FIVE
CHOOSING AND USING PROBIOTICS

When deciding on probiotics, whether in supplement form or found naturally in foods like yogurt and fermented products, it's crucial to consider several factors to maximize their safety and effectiveness. Firstly, choose probiotics that are well-researched and known for their specific strains, as different strains can have varying effects on health. Look for products that disclose the types and amounts of bacteria they contain, ensuring they match your intended health goals.

Moreover, check the expiration date to guarantee potency, as live bacteria are essential for probiotics to be effective.

Storage conditions are also critical; some probiotics require refrigeration to maintain viability. Always follow storage instructions provided by the manufacturer. When starting a probiotic regimen, consult with a healthcare provider, especially if you have underlying health conditions or are pregnant, to ensure they are appropriate for your situation.

Lastly, monitor your body's response. Probiotics affect individuals differently, so pay attention to any changes in digestion or overall well-being. Adjustments in dosage or strain may be necessary based on how your body reacts. By considering these factors, you can make informed choices about probiotics that align with your health needs and promote optimal wellness.

Selecting Probiotic Supplements

Selecting the right probiotic supplements is crucial for ensuring you receive effective support for gut health. Several key criteria should guide your choice to ensure the supplements are both safe and beneficial.

Firstly, consider the strain diversity of the probiotic supplement. Different strains of probiotics offer varying health benefits. Look for products that contain a variety of strains, as this can provide broader support for digestive and immune health. For example, Lactobacillus and Bifidobacterium are common genera of probiotics, each with several strains known for different effects on the body.

Another important factor is the CFU count, which stands for Colony-Forming Units. This indicates the number of viable bacteria present in each dose of the supplement. Higher CFU counts can be advantageous, particularly for therapeutic purposes or when targeting specific health conditions. However, the optimal CFU count can vary depending on individual health needs and the specific strains used.

Survivability of probiotic bacteria is also critical. These bacteria must survive the acidic environment of the stomach to reach the intestines where they exert their beneficial effects. Supplements that utilize enteric coatings or microencapsulation technology are designed to protect probiotics from stomach acid, thereby enhancing their survivability and ensuring they reach the intestines alive.

When choosing probiotic supplements, prioritize those manufactured by reputable brands that adhere to Good Manufacturing Practices (GMP). These standards ensure that the supplements are produced under strict quality control measures. Additionally, opt for supplements that have undergone third-party testing for purity and potency. This independent verification helps confirm that the product contains the ingredients in the amounts stated on the label and is free from contaminants.

Check the expiration date on the supplement packaging to ensure the probiotics are still viable. Probiotics are living organisms, and their effectiveness diminishes over time. Proper storage is also crucial; follow the manufacturer's instructions regarding temperature and storage conditions to maintain the potency of the probiotics until the expiration date.

Selecting high-quality probiotic supplements involves considering strain diversity, CFU count, survivability, quality and manufacturing standards, as well as expiration date and storage conditions. By paying attention to these factors, you can make an informed choice that supports your digestive health effectively. Always consult with a healthcare professional, especially if you have specific health concerns or conditions that may benefit from probiotic supplementation.

Strains and Their Specific Benefits

Different probiotic strains offer a diverse array of health benefits, making it crucial to choose strains that best suit your specific health needs. Understanding the unique properties of each strain can help you optimize the benefits of probiotics for your well-being.

Lactobacillus acidophilus: This strain is renowned for its ability to promote gut health by creating an environment conducive to beneficial bacteria. It aids in digestion by producing lactase, an enzyme that breaks down lactose, which may alleviate symptoms of lactose intolerance such as bloating and gas. By maintaining a healthy balance of gut flora, Lactobacillus acidophilus supports overall digestive health and regularity.

Bifidobacterium bifidum: Primarily found in the large intestine, Bifidobacterium bifidum plays a crucial role in supporting immune function and maintaining gut health. It helps in the digestion of dietary fiber, producing short-chain fatty acids that nourish the colon and contribute to a healthy intestinal environment. Research suggests it may alleviate symptoms of irritable bowel syndrome (IBS) by reducing inflammation and enhancing gut barrier function.

Lactobacillus rhamnosus: Known for its resilience in the digestive tract, Lactobacillus rhamnosus supports both digestive health and immune function. It adheres well to intestinal walls, enhancing its ability to colonize and exert its beneficial effects. This strain is particularly noted for its potential to reduce the duration and severity of diarrhea, including infectious diarrhea and antibiotic-associated diarrhea, by competing with harmful bacteria for nutrients and adhesion sites in the gut.

Saccharomyces boulardii: Unlike bacterial probiotics, Saccharomyces boulardii is a beneficial yeast with unique properties. It has been extensively studied for its ability to prevent and treat various forms of diarrhea, including antibiotic-associated diarrhea and traveler's diarrhea. Saccharomyces boulardii works by inhibiting the growth of harmful bacteria, producing factors that neutralize bacterial toxins, and supporting the restoration of normal bowel function.

Choosing the right probiotic strain depends on your specific health goals. If you seek improved digestion and relief from lactose intolerance symptoms,

Lactobacillus acidophilus may be beneficial. For those aiming to bolster their immune system and manage symptoms of IBS, Bifidobacterium bifidum could be advantageous. Individuals looking to support digestive health and reduce the incidence of diarrhea might consider Lactobacillus rhamnosus or Saccharomyces boulardii, depending on the specific nature of their gastrointestinal concerns.

Incorporating probiotics into your daily regimen can promote a balanced gut microbiota, which is essential for overall health. It's advisable to consult with a healthcare provider or a registered dietitian to determine the most suitable probiotic strains and dosages based on your individual health status and goals. By selecting probiotics tailored to your needs, you can maximize their potential benefits and enhance your overall well-being.

Dosage and Administration

Probiotics have gained popularity for their potential health benefits, particularly in supporting gut health and immune function. However, achieving optimal results from probiotics hinges significantly on how they are dosed and administered. Here's a comprehensive guide on dosage and administration practices to maximize the benefits of probiotics.

Firstly, it's crucial to adhere strictly to the dosage instructions outlined on the supplement label. These guidelines are formulated based on the specific strain(s) of probiotics and their potency, ensuring safe and effective usage. If there are variations in dosage recommendations between products or strains, it's advisable to consult with a healthcare provider for personalized advice.

For individuals new to probiotics, a prudent approach is to start with a lower dose. Beginning with a reduced amount allows the body to acclimate gradually to the introduction of beneficial bacteria. Once accustomed, the dosage can be gradually increased as tolerated, supporting a smoother adjustment period and potentially minimizing any initial digestive discomfort that can sometimes occur.

Timing plays a pivotal role in optimizing probiotic efficacy. In most cases, probiotics are recommended to be taken with meals. This practice helps enhance survivability as the food provides a buffering effect against stomach acids, which can otherwise reduce the viability of probiotic bacteria. However, specific timing instructions may vary by product, and it's advisable to follow manufacturer recommendations for the best outcomes. Consistency is another key principle for reaping the benefits of probiotics. Regular, daily intake helps maintain a steady presence of beneficial bacteria in the gut, supporting ongoing balance and health. Incorporating probiotics into a daily routine, such as taking them at the same time each day, fosters a habit that can promote long-term gut health benefits.

Additionally, the choice of probiotic strains and formulations should align with individual health goals and considerations. Different strains may offer varying benefits, such as supporting digestion, immune function, or alleviating specific digestive issues. Consulting with a healthcare provider can provide insights into which probiotic strains may be most suitable based on personal health needs.

Lastly, monitoring the body's response to probiotics is essential. Individuals may experience diverse reactions, including improvements in digestion, enhanced immunity, or transient digestive adjustments. Keeping track of these responses can guide adjustments in dosage or strain selection to optimize outcomes over time.

Achieving optimal benefits from probiotics requires a thoughtful approach to dosage and administration. By following label instructions, starting slowly, timing intake appropriately, maintaining consistency, and seeking personalized advice when needed, individuals can support their gut health effectively with probiotic supplementation. This proactive approach ensures that probiotics contribute positively to overall well-being and digestive resilience.

Safety and Side Effects

Probiotics, known for their potential health benefits, are generally considered safe for most individuals. However, certain groups and conditions necessitate careful consideration to ensure their safe use:

Pregnant and breastfeeding women should consult healthcare providers before incorporating probiotics into their regimen. While probiotics are generally safe, specific strains and doses may be more suitable during pregnancy or lactation to address individual health needs effectively.

Individuals with compromised immune systems, such as those undergoing chemotherapy or organ transplants, need cautious guidance. Probiotics may pose a risk of infection in immunocompromised individuals, so their use should be strictly supervised by medical professionals.

Potential side effects of probiotics are typically mild and transient. These may include temporary digestive discomfort like gas, bloating, or mild stomach upset. These symptoms often diminish as the body adjusts to the probiotics, but anyone experiencing persistent or severe symptoms should seek medical advice.

Probiotics can interact with medications, especially antibiotics. It's crucial to inform healthcare providers about probiotic use when taking medications to avoid potential interactions that could affect treatment efficacy or safety.

Quality control is paramount when choosing probiotics. Opt for products from reputable brands to minimize the risk of contamination and ensure the product's safety and efficacy. Reliable manufacturers adhere to stringent production standards and provide transparent information about their products.

While probiotics offer potential health advantages, their usage requires careful consideration, particularly in vulnerable populations and when coexisting with certain medical conditions or medications. Pregnant or breastfeeding women, immunocompromised individuals, and those on medications should seek professional guidance before starting probiotics. Monitoring for any adverse effects, even mild ones, is essential, as is selecting products from trustworthy sources to safeguard against potential risks. By taking these precautions, individuals can maximize the benefits of probiotics while minimizing any associated risks to their health and well-being.

CHAPTER SIX

RESEARCH AND ADVANCES IN PROBIOTICS

Probiotics have garnered significant attention in recent years due to their potential health benefits. These live microorganisms, primarily bacteria and some yeasts, are believed to promote gut health and overall well-being when consumed in adequate amounts. Research into probiotics spans various fields, including microbiology, nutrition, and medicine, aiming to understand their mechanisms and potential applications.

One area of focus is the impact of probiotics on gut microbiota. Our intestines host a complex ecosystem of bacteria that play crucial roles in digestion, immune function, and even mental health. Probiotics, such as strains of Lactobacillus and Bifidobacterium, can influence this balance positively by increasing beneficial bacteria and suppressing harmful ones. Studies suggest they may alleviate gastrointestinal disorders like irritable bowel syndrome (IBS) and inflammatory bowel disease (IBD).

Beyond digestive health, probiotics are being explored for their broader systemic effects. They interact with our immune system, potentially enhancing immune responses and reducing inflammation. This has implications for conditions ranging from allergies to autoimmune diseases.

Advancements in probiotic research also extend to mental health. The gut-brain axis, a bidirectional communication network between the gut and the brain, suggests that gut health impacts mood and cognitive function. Probiotics may affect this axis, offering therapeutic potential in managing conditions like depression and anxiety.

Furthermore, the development of personalized probiotics represents a frontier in this field. Tailoring probiotic therapies to an individual's unique microbiome profile could optimize treatment outcomes, ensuring efficacy in diverse populations.

While research continues to expand our understanding, challenges remain.

Determining optimal strains, dosages, and delivery methods (like supplements or foods) are critical for harnessing probiotics' full potential. Regulatory standards also need refinement to ensure product quality and safety.

Ongoing research underscores the promising role of probiotics in enhancing health beyond digestive benefits. As scientists unravel their mechanisms and refine applications, probiotics stand poised to play a pivotal role in future healthcare strategies.

CURRENT RESEARCH

Recent scientific studies are shedding light on the multifaceted benefits of probiotics on gut health and overall well-being. These studies cover a range of topics, from gut microbiota composition to mental and metabolic health, offering promising insights into the potential of probiotics in improving various health conditions.

Gut Microbiota Composition

Recent research has delved into how probiotics influence the gut microbiota's composition and diversity. Specific strains of probiotics have been identified that can restore microbial balance, reduce inflammation, and enhance digestive function. These studies suggest that incorporating probiotics into the diet can significantly alter the gut environment, promoting a healthier microbial community.

Digestive Disorders

Ongoing research is investigating the effectiveness of probiotics in managing various digestive disorders, including irritable bowel syndrome (IBS), inflammatory bowel disease (IBD), and antibiotic-associated diarrhea. Clinical trials have demonstrated that probiotics can alleviate symptoms such as bloating, abdominal pain, and irregular bowel movements, thereby improving the quality of life for individuals suffering from these conditions. The beneficial effects are attributed to probiotics' ability to modulate the gut environment and reduce inflammation.

Immune Function

Probiotics are also being studied for their role in modulating immune responses, particularly in enhancing mucosal immunity in the gut. This line of research indicates that probiotics can reduce susceptibility to infections by strengthening the gut barrier and promoting the production of immune cells. Some studies suggest that regular consumption of probiotics may lower the incidence and severity of respiratory infections and allergies, highlighting their potential as a preventive measure for common illnesses.

Mental Health

The connection between gut health and mental health, known as the gut-brain axis, is an area of growing interest. Research indicates that probiotics may influence neurotransmitter production, reduce inflammation, and alleviate symptoms of anxiety and depression. By modulating the gut microbiota, probiotics can potentially impact brain function and mental well-being, offering a novel approach to managing mental health disorders.

Metabolic Health

Probiotics have been studied for their potential benefits in managing metabolic disorders such as obesity, type 2 diabetes, and cardiovascular diseases. Evidence suggests that probiotics can help regulate glucose metabolism, improve insulin sensitivity, and reduce cholesterol levels. These effects are thought to result from the probiotics' ability to influence gut microbiota composition and function, thereby impacting metabolic processes and promoting overall metabolic health.

The growing body of research highlights the significant potential of probiotics in improving gut health and overall well-being. From restoring microbial balance to enhancing immune function and mental health, probiotics offer a promising avenue for managing various health conditions and improving quality of life.

FUTURE DIRECTIONS

The future of probiotic research is set to delve into several promising trends and areas of exploration, reflecting the rapid advancements in science and technology.

One of the foremost trends is the development of precision probiotics. With significant strides in microbiome sequencing technologies, there is growing potential for personalized probiotic treatments tailored to an individual's unique microbiome profile. This personalized approach aims to enhance therapeutic outcomes by selecting probiotic strains that are best suited to an individual's specific health needs. By tailoring probiotics to the unique composition of a person's gut microbiota, precision probiotics promise to offer more effective and targeted health benefits.

Another emerging trend is the increasing focus on synbiotics, which combine probiotics with prebiotics. Prebiotics are types of dietary fiber that feed the beneficial bacteria in the gut. The synergistic effect of combining these two components is garnering significant attention, as it can enhance gut health more effectively than probiotics or prebiotics alone. Future research is likely to concentrate on developing optimized synbiotic formulations designed to deliver specific health benefits. These tailored formulations could address various health issues more precisely by promoting a more favorable gut environment for beneficial bacteria.

Mechanistic studies represent another critical area of future research. A deeper understanding of the precise mechanisms through which probiotics exert their beneficial effects is essential for advancing the field. This includes investigating the interactions between probiotics and host cells, as well as exploring microbial signaling within the gut environment. By elucidating these mechanisms, researchers can better understand how probiotics influence the gut microbiota, immune system, and other physiological pathways. This knowledge will not only advance scientific understanding but also inform the development of more effective probiotic therapies.

The integration of probiotics into clinical practice is another important direction for future research. Developing evidence-based guidelines for healthcare providers is crucial for the effective use of probiotics in treating specific health conditions. Future research will likely focus on defining the optimal probiotic strains, dosages, and treatment durations for various health issues. By establishing these guidelines, probiotics can be more reliably incorporated into treatment protocols, improving patient outcomes.

The future of probiotic research is poised to explore precision probiotics, synbiotics, mechanistic studies, and clinical applications. These emerging trends reflect a broader shift towards personalized, evidence-based approaches in healthcare, leveraging the power of the microbiome to improve health outcomes. Through these efforts, probiotics are set to play an increasingly significant role in promoting health and preventing disease.

PERSONALIZED NUTRITION AND PROBIOTICS

Personalized nutrition focuses on tailoring dietary recommendations to an individual's unique genetic profile, lifestyle choices, and overall health status. Within the scope of probiotics, this personalized approach encompasses several key strategies aimed at enhancing health outcomes through targeted interventions.

One of the primary strategies is microbiome profiling. This involves conducting detailed analyses of an individual's gut microbiota to identify specific imbalances and deficiencies. By examining the composition and diversity of the gut microbiome, healthcare providers can pinpoint which beneficial bacteria are lacking or which harmful microbes are overrepresented. This information is crucial in designing probiotic treatments that are precisely aligned with the individual's microbial needs. For instance, if a person's gut shows a deficiency in Lactobacillus, a probiotic strain known for its benefits to gut health, a personalized probiotic regimen may include supplements rich in this specific strain to restore balance and promote a healthy microbiome.

Another important aspect of personalized probiotics is targeted therapies. These therapies are customized to address specific health conditions or symptoms that are identified through comprehensive health assessments. For example, an individual suffering from irritable bowel syndrome (IBS) may benefit from a specific blend of probiotics that have been shown to alleviate IBS symptoms. By tailoring probiotic treatments to the individual's specific health issues, the likelihood of achieving positive health outcomes is significantly increased. This approach contrasts with the one-size-fits-all probiotic recommendations that are commonly seen, offering a more effective and precise intervention.

Response prediction is another crucial element of personalized probiotics. By analyzing an individual's microbiome composition and genetic factors, it is possible to predict how they might respond to different probiotic strains. This predictive capability allows for the optimization of probiotic treatments, ensuring that individuals receive the strains that are most likely to produce beneficial effects for them. For instance, genetic markers might indicate a predisposition to certain digestive issues that could be mitigated by specific probiotics, allowing for preemptive and targeted intervention.

The integration of personalized nutrition and probiotics represents a sophisticated approach to health and wellness. Through microbiome profiling, targeted therapies, and response prediction, personalized probiotic interventions can be tailored to meet the unique needs of each individual. This personalized approach not only enhances the effectiveness of probiotic treatments but also aligns with the broader trend towards individualized healthcare, where treatments and recommendations are customized to optimize individual health outcomes. By leveraging these advanced strategies, personalized probiotics hold the promise of more precise and effective management of various health conditions, ultimately contributing to better overall health and well-being.

PROBIOTICS IN CLINICAL PRACTICE

In clinical practice, healthcare providers increasingly recognize the therapeutic potential of probiotics, incorporating them into treatment plans for a variety of conditions.

Gastroenterology: Probiotics are frequently recommended for managing digestive disorders such as Irritable Bowel Syndrome (IBS), Inflammatory

Bowel Disease (IBD), and antibiotic-associated diarrhea. These beneficial bacteria may be used alongside conventional treatments to enhance symptom relief and decrease the frequency of disease flare-ups. For instance, in patients with IBS, probiotics can help reduce bloating, abdominal pain, and irregular bowel movements. In cases of IBD, probiotics may support maintaining remission and improving the gut's mucosal barrier function, thus potentially preventing relapses.

Pediatrics: In pediatric health, probiotics have shown significant benefits, particularly in reducing the incidence of infantile colic, eczema, and respiratory infections. Pediatricians might recommend probiotics to support both immune development and digestive health in children. For infants with colic, specific probiotic strains can help alleviate crying and discomfort. Probiotics are also linked to lower rates of eczema, a common inflammatory skin condition in children, by potentially modulating the immune response. Additionally, probiotics can enhance respiratory health, reducing the frequency and severity of infections, which is crucial for young children with developing immune systems.

Obstetrics and Gynecology: Probiotics are being explored for their role in maintaining vaginal health and preventing urinary tract infections

(UTIs) in women. They may be particularly beneficial during pregnancy, supporting both maternal and fetal health. Probiotics can help maintain a healthy vaginal microbiome, which is essential for preventing infections and complications during pregnancy. They are also considered for their potential to reduce the risk of UTIs by inhibiting the growth of harmful bacteria in the urinary tract. By maintaining a balanced microbiome, probiotics contribute to overall reproductive health.

General Practice: In primary care settings, healthcare providers often recommend probiotics to support overall gut health, immune function, and mental well-being. The gut microbiome plays a crucial role in these areas, and probiotics can help maintain its balance. Providers can guide patients in selecting appropriate probiotic supplements and incorporating probiotic-rich foods, such as yogurt, kefir, and fermented vegetables, into their diets. This holistic approach not only aids in preventing and managing specific health conditions but also promotes overall wellness by enhancing the body's natural defenses and improving mental health, potentially through the gut-brain axis.

Overall, the integration of probiotics into clinical practice represents a growing recognition of their diverse benefits. Healthcare providers across various specialties are utilizing probiotics to enhance traditional treatment modalities, promoting better health outcomes for their patients.

☐

CHAPTER SEVEN

LIFESTYLE FACTORS AND GUT HEALTH

Gut health is influenced not just by diet and probiotics but also by various lifestyle factors. These include stress, exercise, sleep, and environmental exposures. Here's how these factors affect the gut microbiome and strategies to manage their impact:

Stress and Gut Health

Stress has a profound impact on gut health, primarily through the gut-brain axis—a communication network connecting the central nervous system with the enteric nervous system, which regulates gastrointestinal function. When stress occurs, this communication can be disrupted, leading to various adverse effects on the gut.

One significant impact of stress is altered gut motility. Stress can influence the way the intestines contract, often resulting in symptoms such as diarrhea or constipation. This is due to the fact that stress hormones, such as cortisol, can interfere with the normal rhythmic contractions of the gut, leading to either accelerated or slowed transit times.

Chronic stress can also imbalance the gut microbiota. The microbiota consists of a complex community of microorganisms that play a crucial role in maintaining gut health. Under stress, the composition of these microorganisms can change, reducing the population of beneficial bacteria while promoting the growth of harmful ones. This imbalance can compromise digestive health and is linked to conditions such as irritable bowel syndrome (IBS) and inflammatory bowel disease (IBD).

Moreover, stress can increase intestinal permeability, a condition often referred to as "leaky gut." Normally, the intestinal barrier tightly regulates what passes into the bloodstream. However, stress can weaken this barrier, making it more permeable. This allows harmful substances like toxins, pathogens, and undigested food particles to enter the bloodstream, potentially triggering inflammation and immune responses throughout the body. This increased permeability is associated with various health issues, including autoimmune diseases and chronic inflammatory conditions.

Strategies to Manage Stress:

Managing stress effectively involves incorporating several key strategies into your daily routine, each contributing to your overall well-being and reducing the impact of stress on your life.

Mindfulness and Meditation: Engaging in mindfulness practices, such as meditation, can significantly reduce stress levels. These practices involve focusing on the present moment and acknowledging your thoughts and feelings without judgment. Regular mindfulness meditation promotes relaxation, helps clear the mind, and can lead to a more balanced emotional state, making it easier to handle stress.

Exercise: Regular physical activity is a powerful tool for managing stress. Exercise releases endorphins, which are natural mood lifters, and helps lower levels of stress hormones like cortisol. Additionally, physical activity benefits gut health by promoting the growth of beneficial gut bacteria, which can enhance your overall mood and reduce anxiety. Incorporating activities like walking, running, yoga, or any form of exercise you enjoy can significantly improve your stress management.

Healthy Diet: Nutrition plays a crucial role in managing stress. A balanced diet rich in fruits, vegetables, and whole grains provides essential nutrients that support brain function and stabilize your mood. Foods high in antioxidants, omega-3 fatty acids, and probiotics can help mitigate the negative effects of stress on the gut and overall health. Avoiding excessive caffeine, sugar, and processed foods can also help maintain stable energy levels and reduce stress.

Adequate Sleep: Getting enough restful sleep is essential for stress management and overall health. Poor sleep can exacerbate stress, while adequate sleep enhances your ability to cope with stressors. Establishing a regular sleep schedule, creating a relaxing bedtime routine, and ensuring your sleep environment is conducive to rest can improve sleep quality. Aim for 7-9 hours of sleep per night to support emotional stability and physical health.

Exercise and the Gut

Engaging in regular physical activity yields numerous benefits for gut health, significantly influencing the gut microbiome's composition and function. Exercise promotes a more diverse gut microbiota, a key indicator of overall health. Greater microbial diversity is linked to improved immune function, enhanced digestion, and a reduced risk of gastrointestinal disorders. By fostering a varied microbial environment, exercise helps maintain a balanced gut ecosystem, which is crucial for optimal health.

Physical activity also strengthens the intestinal barrier, enhancing gut barrier function. This barrier is vital in preventing harmful substances from entering the bloodstream, thereby reducing the risk of a condition commonly referred to as "leaky gut." A strong intestinal barrier ensures that the gut can efficiently filter out toxins and pathogens while allowing essential nutrients to pass through. This protective effect of exercise on the gut barrier can help prevent inflammatory responses and maintain overall gut integrity.

Moreover, regular exercise improves bowel movement regularity, positively impacting gut transit time. Efficient transit time means that food moves through the digestive tract at an optimal pace, which can prevent constipation and reduce the risk of developing gastrointestinal issues. Improved transit time also supports the growth of beneficial bacteria in the gut, as it minimizes the duration that waste products stay in the intestines.

The interplay between physical activity and gut health extends to metabolic benefits as well. Exercise influences the production of short-chain fatty acids (SCFAs), which are crucial for gut health and play a role in regulating inflammation, boosting immunity, and providing energy for colon cells. By enhancing SCFA production, exercise further supports a healthy gut environment.

Sleep and Gut Health

Quality sleep plays a crucial role in maintaining a healthy gut microbiome. A lack of sufficient, restful sleep can disrupt the delicate balance of gut bacteria, leading to a range of health issues.

One of the primary consequences of poor sleep is the alteration in microbiota composition. When sleep is disrupted, the diversity and abundance of beneficial bacteria in the gut decrease, while harmful bacteria proliferate. This imbalance can compromise gut health and overall well-being. For instance, studies have shown that sleep deprivation can significantly reduce populations of beneficial bacteria like Lactobacillus and Bifidobacterium, which are essential for digestion and immunity. Concurrently, there is often an increase in pathogenic bacteria, contributing to a less favorable gut environment.

Moreover, inadequate sleep can lead to systemic inflammation, which further impacts gut health negatively. Sleep deprivation triggers an inflammatory response in the body, characterized by increased levels of pro-inflammatory cytokines. This chronic inflammation can damage the gut lining, making it more permeable. This condition, often referred to as "leaky gut," allows harmful substances to pass through the gut wall into the bloodstream, potentially leading to various inflammatory diseases and conditions. Hormonal imbalances are another significant effect of poor sleep on gut health. The hormones ghrelin and leptin, which regulate appetite and satiety, are particularly affected by sleep patterns. Lack of sleep leads to elevated ghrelin levels, which increase appetite, and reduced leptin levels, which diminish feelings of fullness. This hormonal imbalance can result in overeating and poor dietary choices, further exacerbating gut health issues. Additionally, sleep deprivation affects the production of cortisol, the stress hormone, which can alter gut motility and increase gut permeability.

Tips for Better Sleep:

Ensuring quality sleep is crucial for overall health and well-being. Here are some effective tips to improve your sleep:

Firstly, maintaining a regular sleep schedule is key. Going to bed and waking up at consistent times every day helps regulate your body's internal clock, making it easier to fall asleep and wake up refreshed. Even on weekends, try to stick to your schedule to avoid disrupting your sleep patterns.

Secondly, creating a restful sleep environment is essential. Your bedroom should be conducive to sleep—dark, quiet, and cool. Consider using blackout curtains or an eye mask to block out light, earplugs or a white noise machine to mask sounds, and adjusting the temperature to a comfortable level, usually cooler temperatures promote better sleep.

Thirdly, limit screen time before bed. The blue light emitted by smartphones, tablets, and computers can interfere with melatonin production, making it harder to fall asleep. Try to avoid screens for at least an hour before bed, or use blue light filters and night mode settings on devices to minimize the impact.

Lastly, practicing relaxation techniques can help signal to your body that it's time to wind down. Activities like reading a book, taking a warm bath, practicing gentle yoga, or engaging in meditation can all promote relaxation and reduce stress levels before bedtime. Developing a bedtime routine that includes these calming activities can prepare your mind and body for sleep.

Incorporating these tips into your daily routine can significantly improve the quality of your sleep over time. Remember that everyone's sleep needs are different, so it may take some experimentation to find the combination of strategies that work best for you. Prioritizing good sleep hygiene is an investment in your health, mood, and overall productivity throughout the day.

Environmental Factors

Environmental factors play a significant role in influencing gut health, with exposures like antibiotics and pollutants exerting notable impacts:

Antibiotics are indispensable in combating bacterial infections; however, their use can disrupt the delicate balance of the gut microbiota. This disruption, known as dysbiosis, occurs because antibiotics do not discriminate between harmful and beneficial bacteria, leading to a reduction in diversity and abundance of beneficial microbes. Such alterations can predispose individuals to infections like Clostridium difficile, a bacterium that opportunistically thrives in a depleted gut environment. Moreover, repeated or prolonged antibiotic treatments may exacerbate these effects, potentially impairing immune function and metabolic processes regulated by the gut microbiota.

Pollutants, encompassing substances such as heavy metals, pesticides, and industrial chemicals, also pose significant threats to gut health. These contaminants enter the body through various pathways including ingestion, inhalation, and dermal absorption. Once inside, pollutants can disrupt the gut microbiota composition by selectively promoting the growth of certain bacteria while diminishing others. This imbalance not only compromises the microbiota's ability to perform essential functions like nutrient absorption and immune modulation but also triggers inflammatory responses within the gastrointestinal tract. Chronic exposure to pollutants has been linked to persistent inflammation, which can contribute to the development of gastrointestinal disorders and other systemic health problems over time.

Understanding the implications of these environmental factors on gut health underscores the importance of proactive measures to mitigate their effects.

Strategies may include judicious use of antibiotics under medical supervision, alongside efforts to support gut microbiota resilience through probiotics, prebiotics, and dietary adjustments. Similarly, reducing exposure to pollutants through environmental regulations and personal lifestyle choices can help safeguard gut health and overall well-being. By addressing these factors comprehensively, we can promote a healthier balance within the gut microbiota and potentially reduce the incidence of associated health complications in the long term.

☐

Strategies to Mitigate Environmental Impacts:

Mitigating environmental impacts necessitates adopting conscientious practices across various facets of daily life. One crucial strategy involves the judicious use of antibiotics, reserving them solely for instances where prescribed by healthcare professionals. This measure not only curbs the development of antibiotic-resistant bacteria but also minimizes the release of pharmaceutical residues into the environment, which can adversely affect aquatic ecosystems and potentially contribute to antibiotic resistance in environmental bacteria. Additionally, post-antibiotic probiotic supplementation aids in restoring gut microbiota balance, enhancing individual health outcomes while indirectly reducing environmental stressors.

Reducing exposure to pollutants constitutes another pivotal tactic. Opting for natural cleaning products diminishes the introduction of harmful chemicals into waterways during disposal, thereby safeguarding aquatic biodiversity. Choosing organic produce whenever feasible reduces agricultural runoff of synthetic pesticides and fertilizers, preserving soil health and minimizing contamination of groundwater sources. Moreover, abstaining from smoking and moderating alcohol consumption not only promotes personal health but also mitigates air and water pollution associated with tobacco cultivation and alcohol production processes.

Supporting natural detoxification processes through dietary choices reinforces the body's innate mechanisms for eliminating toxins. A diet rich in fiber aids in binding and excreting harmful substances from the gastrointestinal tract, while antioxidants combat oxidative stress induced by environmental pollutants. Consuming nutrient-dense foods further bolsters overall health, reducing the likelihood of chronic diseases that can burden healthcare systems and exacerbate environmental impacts through medical waste and pharmaceutical runoff.

Adopting these strategies—mindful antibiotic use, pollutant reduction, and dietary support for detoxification—constitutes a proactive approach to mitigating environmental impacts. By intertwining individual health with environmental stewardship, these practices collectively contribute to a sustainable future where both human and ecological well-being thrive harmoniously.

THE END

www.ingramcontent.com/pod-product-compliance
Lightning Source LLC
Chambersburg PA
CBHW071830210526
45479CB00001B/68